NURSE SHARKS

by Julie K. Lundgren

A Crabtree Seedlings Book

TABLE OF CONTENTS

Crabtree Publishing

crabtreebooks.com

School-to-Home Support for Caregivers and Teachers

This book helps children grow by letting them practice reading. Here are a few guiding questions to help the reader with building his or her comprehension skills. Possible answers appear here in red.

Before Reading:

- What do I think this book is about?
  - *I think this book is about nurse sharks.*
  - *I think this book is about how gentle and caring they are.*
- What do I want to learn about this topic?
  - *I want to learn about the habits of nurse sharks.*
  - *I want to learn where nurse sharks live.*

During Reading:

- I wonder why...
  - *I wonder why they are called nurse sharks.*
  - *I wonder why nurse sharks rest during the day.*
- What have I learned so far?
  - *I have learned that nurse sharks hide under reefs.*
  - *I have learned that they hunt during the night.*

After Reading:

- What details did I learn about this topic?
  - *I have learned that nurse sharks can rest but great white sharks can't rest because they must swim to breathe.*
  - *I have learned that nurse sharks suck up fish, shrimp, and squid.*
- Read the book again and look for the vocabulary words.
  - *I see the word **reef** on page 3 and the word **barbels** on page 16. The other glossary words are on pages 22 and 23.*

NURSE SHARKS

What hides under the **reef**?

Nurse sharks! During the day, they rest.

Other sharks cannot rest.
They must swim to breathe.

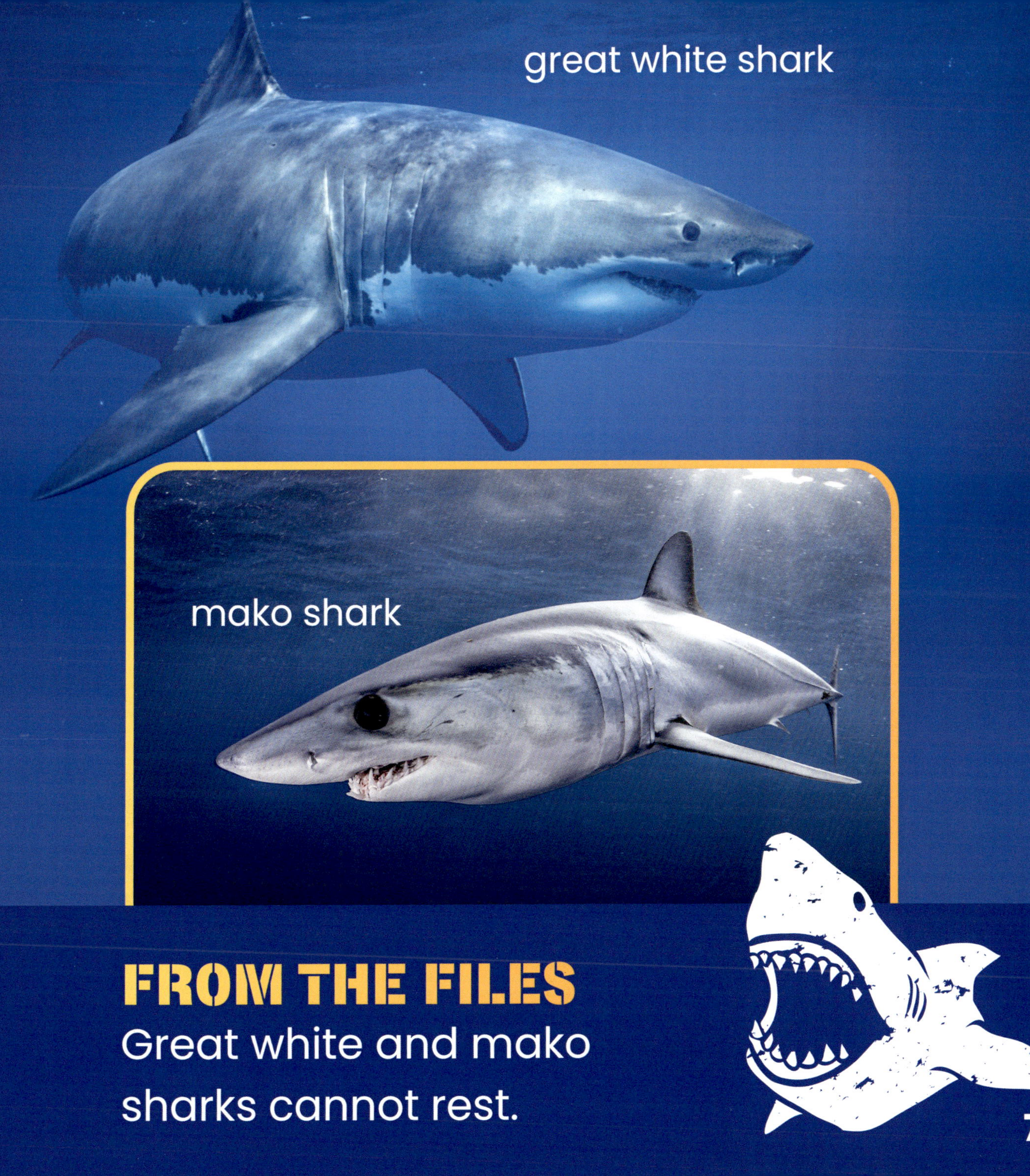

FROM THE FILES

Great white and mako sharks cannot rest.

At night, nurse sharks hunt.

They do not chase their **prey**.

Instead, they slowly roam the seafloor.

They suck up fish, shrimp, and squid.

squid

FROM THE FILES

They crunch lobsters, too.

Barbels help them find prey by touch.

FROM THE FILES

Catfish use barbels the same way.

Divers spot them more than any other shark.

FROM THE FILES

You might spot one in an **aquarium**!

They are gentle, but do not pet one!

FROM THE FILES

They have lots of sharp **teeth**!

GLOSSARY

aquarium (uh-KWAIR-ee-um): An aquarium is a place people can visit to see fish and other water animals.

barbels (BAR-bulls): Barbels are like long, soft fingers on a nurse shark's face that help find food by touch

divers (DIE-verz): Divers are people that wear gear to help them breathe underwater.

prey (PRAY): Prey is any animal hunted and eaten by another animal.

reef (REEF): A reef is a shallow, underwater ridge where many sea animals live.

teeth (TEETH): Teeth are white, bony parts inside a mouth that are used for biting and chewing.

Index

Websites

https://aqua.org/explore/animals/nurse-shark

www.montereybayaquarium.org/animals/animals-a-to-z/sharks

About the Author

Julie K. Lundgren

Julie K. Lundgren grew up near Lake Superior where she reveled in mucking about in the woods, picking berries, and expanding her rock collection. Her interests led her to a degree in biology. She lives in Minnesota with her family.

Crabtree Publishing

crabtreebooks.com 800-387-7650

Written by: Julie K. Lundgren
Designed by: Jennifer Dydyk
Edited by: Kelli Hicks
Proofreader: Melissa Boyce

Photographs:
Shark illustration on cover logo © BATKA/Shutterstock; white shark illustration for "FROM THE FILES" © Dashikka/Shutterstock; Cover © Carlos Grillo/Shutterstock.com; page 3 © Richard Whitcombe/Shutterstock.com; page 5 © Carlos Aguilera/Shutterstock.com; page 7 great white © KDR In-Focus Productions/Shutterstock.com; mako © wildestanimal/ Shutterstock.com; page 9 © nicolasvoisin44/Shutterstock.com; page 11 © Eric Carlander/Shutterstock.com; page 13 © Keith Levit/Shutterstock.com; page 15 squid © George P Gross/Shutterstock.com, lobster © MIGUEL G. SAAVEDRA/Shutterstock.com; page 17 nurse shark © Yann hubert/Shutterstock.com, catfish © KT photo/Shutterstock.com; page 19 diver © Jag_cz/Shutterstock.com, aquarium © Evikka/Shutterstock.com; page 21 © frantisekhojdysz/Shutterstock.com;

Hardcover 978-1-4271-5854-3
Paperback 978-1-4271-5855-0
Ebook (pdf) 978-1-4271-5856-7
Epub 978-1-4271-5857-4
Read-along 978-1-4271-5858-1
Audio book 978-1-4271-5859-8

Printed in Canada/092023/CPC20230926

Published in Canada
Crabtree Publishing
616 Welland Avenue
St. Catharines, Ontario
L2M 5V6

Published in the United States
Crabtree Publishing
347 Fifth Avenue
Suite 1402-145
New York, NY 10016

Library and Archives Canada Cataloguing in Publication

Available at the Library and Archives Canada

Library of Congress Cataloging-in-Publication Data

Names: Lundgren, Julie K., author.
Title: Nurse sharks / Julie K. Lundgren.
Description: New York : Crabtree Publishing, [2022] | Series: Shark files - a Crabtree seedlings book | Includes index.
Identifiers: LCCN 2021018450 (print) | LCCN 2021018451 (ebook) | ISBN 9781427158543 (hardcover) | ISBN 9781427158550 (paperback) | ISBN 9781427158567 (ebook) | ISBN 9781427158574 (epub) | ISBN 9781427158581
Subjects: LCSH: Nurse shark--Juvenile literature.
Classification: LCC QL638.95.G55 L86 2022 (print) | LCC QL638.95.G55 (ebook) | DDC 597.3/3--dc23
LC record available at https://lccn.loc.gov/2021018450
LC ebook record available at https://lccn.loc.gov/2021018451